Wheels

Diggers & Loaders

Graham Thompson

Macdonald

Digger Loader

This vehicle has a back hoe and a loader all in one. The driver has two sets of controls, one for the back hoe and one for the loader. The hoe is used to dig up earth and scrape it into a big pile. The loader scoops up the pile to carry it away.

Bulldozer

The bulldozer is big and powerful. It moves along on caterpillar tracks so it can drive over the roughest ground. In front it has a great dozer blade. This pushes away piles of rocks and earth to make the ground flat.

Tip-up truck

This tip-up truck drives along the road to the building site. It carries a heavy load of rocks to fill up a pit in the ground. When the driver gets there, he works the controls so that the back of the truck tips up. With a loud rumble all the rocks roll into the pit.

Crane

This crane is lifting crates off a ship. It is easy to lift very heavy loads like this. The driver sits in his cab and works the long lifting arm. It lifts up its load and swings it round on to a waiting truck. The crane will not tip over because its own truck is very heavy.

Dumper truck

Dumper trucks are small but tough.
They are used to carry heavy loads and have
thick tyres for driving on rough ground.

These dumper trucks are bringing cement to fill up a hole in the road. Up tips the front and down pours the cement into the hole. And here comes another load!

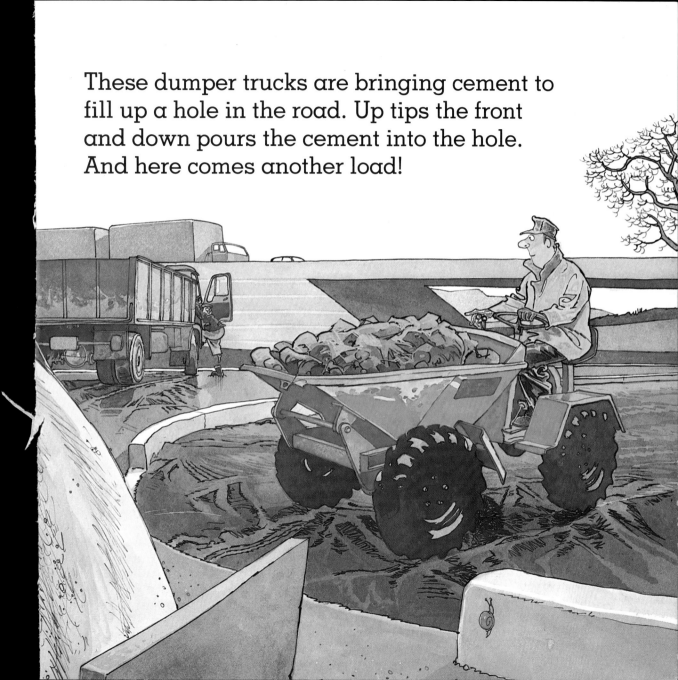

Road roller

The road roller moves slowly along. Its big heavy wheels roll round grinding and pressing down the hot, sticky tar to make the new road smooth and flat. Once the tar has dried hard, it will be ready for cars and trucks to drive along.

Elevated platform truck

At the airport an engineer is lifted high up alongside the aeroplane. Now he can clean the windscreen and check the lights before the plane takes off again. When he has finished he pulls a lever and the long extended arm folds down on to the truck.

Mini transporter

This tough little transporter can travel up the steepest hillside. It has a lot of strong wheels so it can drive across rough ground. In winter the farmer uses it to bring food to his sheep from the farm in the valley.

Tree harvester

This truck works in the forest. It moves across the muddy ground on caterpillar tracks. The driver sits safe in his cabin and works the controls. The tree harvester clamps shut on the tree and saws right through the trunk. Then it lifts it up into the air and the driver steers the truck across the forest-clearing to the wood pile.